UNIVERSITY OF IDAHO
AGRICULTURAL EXPERIMENT STATION
Department of Agricultural Engineering

The Measurement of Water

A HAND BOOK FOR DITCH RIDERS AND WATER USERS

By

W. G. STEWARD

BULLETIN NO 127 JANUARY, 1922

Published by the University of Idaho, Moscow, Idaho

UNIVERSITY OF IDAHO AGRICULTURAL EXPERIMENT STATION

BOARD OF REGENTS

EXECUTIVE COMMITTEE

EXPERIMENT STATION STAFF

THE MEASUREMENT OF WATER

W. G. STEWARD

Importance of Water Measurements

"The measurement of water is the great irrigation need of the day, in the face of which all other needs vanish."*

The importance of water measurement in the state of Idaho is very great. Comparatively few of the irrigation companies in the state have any systematic method of making measurements, or of keeping records of water delivered to the users. The matter of equitable distribution and accounting is so important that it should command the universal attention of irrigators throughout the state.

What farmer in Idaho would sell 1000 bushels of wheat to ten of his neighbors and as each came for his wheat, would say "It is out in the granary, go and help yourself to what you think is your share. There is no need to measure or weigh it, just guess at it as closely as you can." How much wheat do you think would be left for the last two or three men? In the distribution of water in the state of Idaho a not uncommon practice has been: "Help yourself." Lax methods of measurement and distribution often led to disputes and expensive litigation.

The object of this bulletin is not so much to give new information as to give to the farmers and ditch riders the simplest known methods of water measurement with as much of the practical as possible and with only as much theory as is necessary. No attempt has been made to hold to any of the refinements such as might be used by the engineer in scientific investigations; at the same time it is intended to keep well within the limits of the needed accuracy. It is not the intention to encourage lax or careless methods, but to encourage a more universal practice of water measurement by furnishing the necessary information in the simplest terms possible.

It is the desire of the author to present for the benefit of the irrigators of Idaho the best practice and methods evolved from years of experimenting with practically every style of measuring device ever put on the market. It is believed that the information here given will enable a ditch rider to make a close approximation of the flow of water under almost any condition that may occur on an irrigation project.

* John A. Widtsoe in Principles of Irrigation Practice.

Units of Measurement

For the measurement of water there are two kinds or classes of units. These are:

1. The units of *quantity* or *volume*, the most common of which are the gallon, the cubic foot, and the acre foot.

2. The units of the *rate of flow* which take into account the time consumed in the delivery of a unit of volume. The most common of these are the miner's inch, gallons per minute, cubic feet per second, and acre foot per day.

It is important to get well fixed in mind the distinction between a unit of volume and a unit of rate of flow; for instance, a cubic foot is a definite quantity of water such as would be contained in a cubical can having faces one foot square. A cubic foot per second, or what is known as a second foot, would fill such a can once every second as long as it continued to flow, that is, 60 cubic feet per minute or in one hour a flow of a second foot would have discharged 3600 cubic feet of water.

The Second Foot or the Cubic Foot Per Second.

The legislature of the State of Idaho in 1899 passed a law making the cubic foot per second the legal unit for the measurement of irrigation water. A cubic foot per second flowing for twenty-four hours would deliver 86,400 cubic feet per day and as there are 43,560 square feet in an acre this amount of water would cover 1 acre 1.9835 feet deep, or would equal 1.9835 acre feet. This is so near the 2 acre feet that it is customary to say that one second foot of water flowing for one day equals 2 acre feet.

Acre Foot.

An acre foot of water is the quantity of water that would cover one acre to a depth of one foot. As there are 43,560 square feet in one acre, it is equal to 43,560 cubic feet.

The Miner's Inch.

The miner's inch in Idaho is the one-fiftieth part of a second foot. The term miner's inch came from the fact that water was formerly measured by the amount of water passing thru an orifice one inch square under a given pressure. The miner's inch in Idaho was the water that would pass thru a sharp-edged orifice one inch square under a pressure of 4 inches measured from the center of the orifice. This is equal to approximately one-fiftieth of a second foot.

Orifice.

An orifice is an opening in a heading, preferably of rectangular shape, through which water flows under pressure. The standard orifice has sharp edges so that the jet of water passing thru is contracted on all four sides.

If the discharge of the water thru the orifice is into free air, it is called a "Free Orifice." If the discharge of water is under water, it is called a "Submerged Orifice." The head used in computing the discharge of a free orifice is the distance from the surface of the water above the orifice down to the center of the orifice. The head used in computing the discharge of a submerged orifice is the difference in elevation between the water surface above the orifice and the water surface below the orifice. The same discharge table is used in each case.

Head.

The term head is used in two senses in irrigation (1) meaning the quantity of water available, as when one says "a large head of water," and (2) meaning the height of the water surface above the weir crest or the difference in the level of the water above and below a submerged orifice.

Weir Scale or Gage.

A weir scale is a strip of metal or other material marked with some standard measuring units, for instance, into feet, tenths of a foot, and hundredths of a foot. It is used to measure the depth of water flowing over the weir crest. A weir scale may be marked into feet and inches, but the decimal marking is usually considered more convenient.

Check Board.

A check board is a board placed in a canal structure to check the flow of water, causing the water level to be raised. A check board acts as a weir when water is flowing over it.

Heading.

A canal heading is any device which may be used for changing the flow of water from one channel to another.

Methods of Measurement

Irrigation water is usually measured by one of three methods: over a weir, thru an orifice, or in an open channel. Of these methods, the one most commonly used by ditch riders is the weir.

A weir is a notch in a vertical wall through which water flows. The weir is the best instrument ever devised for common use in the measurement of irrigation water. It is cheap and simple of construction. The results are accurate and easily understood. The measurements are easy to make and the computations are rapid.

There are many types of weirs, but the one most commonly used in the measurement of irrigation water is the Cippoletti weir, so called be-

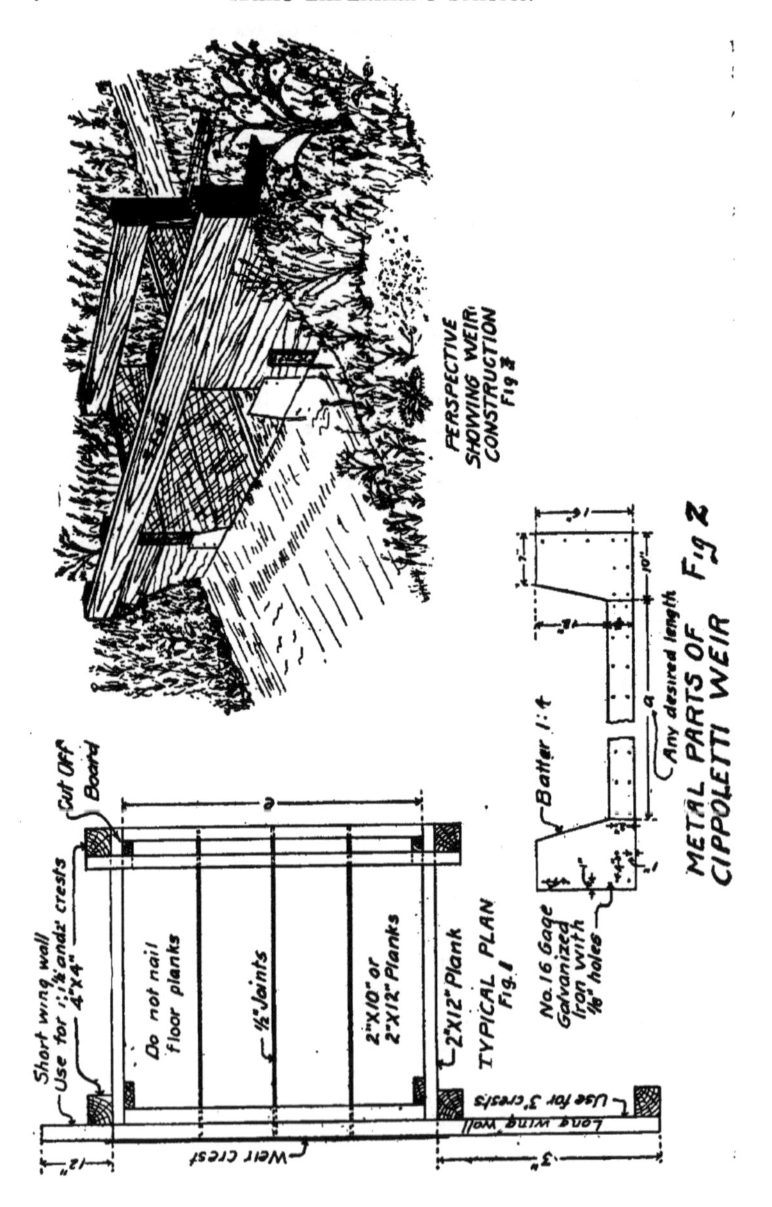

TYPICAL PLAN Fig. 1

METAL PARTS OF CIPPOLETTI WEIR Fig 2

PERSPECTIVE SHOWING WEIR CONSTRUCTION Fig 3

cause it was designed by an Italian engineer named Cesare Cippoletti. This weir has a thin horizontal crest and thin sides, and the weir notch is wider across the top than at the bottom,*the sides having a slope of one inch out to four inches rise or what is usually termed a 1:4 slope*.

A rectangular suppressed weir has a right angle notch, the sides of which are flush with the weir box in which it is placed, so that the water flowing over the crest is not contracted on the sides. (see Figure 6, showing types of weirs). The suppressed rectangular weir discharge is usually computed from the Cippoletti weir tables.

The rectangular contracted weir has a rectangular notch, the sides of which are some distance from the weir box so that the water flowing over the crest is contracted on the sides as in the case of a Cippoletti weir. The rectangular contracted weir requires a special table for each length of weir and for this reason is not generally used. Where the width of the crest is large compared with the depth, a Cippoletti table can be used without appreciable error. This condition obtains in most cases where water runs over checks and waste gates. The Cippoletti weir and the suppressed rectangular weir discharges are usually computed from the same table.

Construction of Weirs

The construction of a weir box and crest is shown in the accompanying drawings, of which Figure 1 is a typical plan, the view showing two types of wings. The first is for the smaller sized weirs with 1, 1½, and 2 foot crests. The wings for this size are formed by a 1 inch by 12 inch plank nailed vertically against the upright post as shown in Figures 1 and 5. The other type of wing is for crests of 3 feet and over. The wing can be made longer or shorter than shown according to the type of soil in which the weir structure is set. Figure 5 shows how the weir crest is mounted on a weir board and how this board is mounted in front of a flush board so that it can be raised or lowered to an amount equal to the approximate width of the flush board without making a break in the face of the structure. This is to allow the weir crest to be adjusted to meet the requirements of the varying change in the water level in the canal without otherwise disturbing the weir structure. In order to facilitate any such necessary change the weir board should be so nailed in place that it can be easily removed and replaced at any desired height.

Figure 5 shows how the sides and bottom of the weir box are placed and also the method of mounting the metal crest on the weir board. It will be noted that the blade or weir proper is composed of three pieces of sheet metal, (see Figure 2), the crest and two side pieces. They are so made as to allow the use of any width of crest with the same size of sides. Figures 3 shows a weir box set in place with the weir and gage

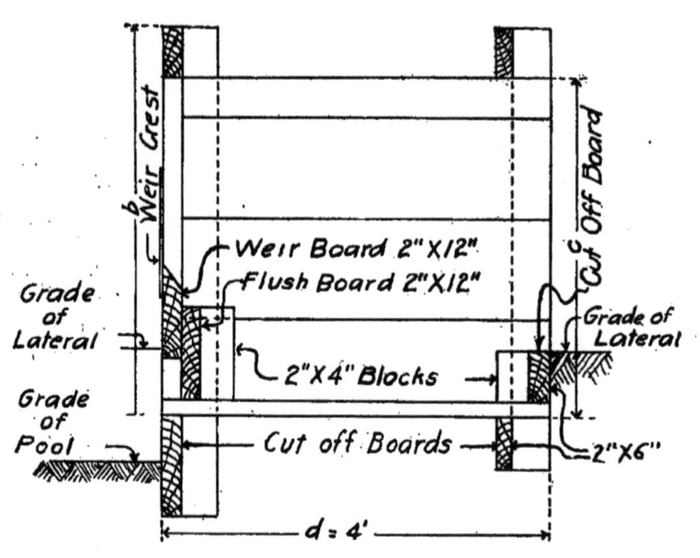

TYPICAL LONGITUDINAL SECTION
Fig 4

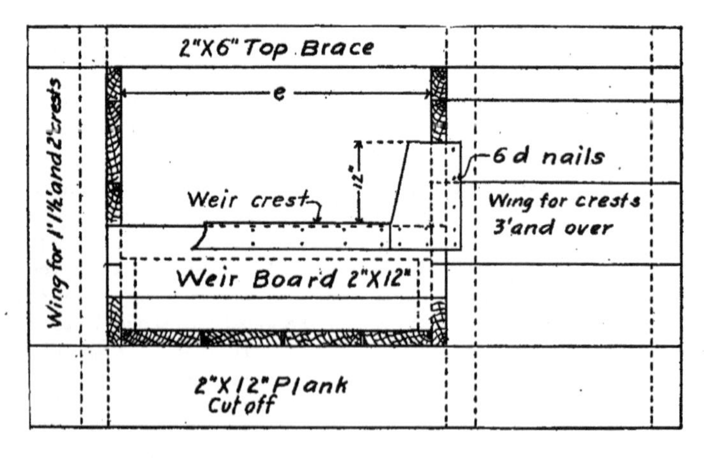

TYPICAL UPSTREAM ELEVATION
Fig. 5

mounted thereon and gives a good idea of the comparative width of weir pool required. Figure 2 gives the dimensions of the metal parts of a weir with the location of the nail holes required to fasten them in place.

The depression in the floor as shown in the drawings may be omitted and the floor may be set even with the canal grade if the soil does not wash easily or if riprap can be easily and cheaply placed below the weir box to prevent the soil erosion. This permits reducing the height of the box. Other changes as to the width of the wings, the depth of the cut-off board and the height of the flush board may be made as conditions demand.

The requirements for the proper setting and operating of this type of weir are:

1. It should be set at the lower end of a long pool sufficiently wide and deep to give an even smooth current with a velocity of approach of not over 0.5 of a foot per second, which means practically still water.

2. The line of the weir box should be parallel with the direction of flow, that is, the crest is to be at right angles to the direction of the flow.

3. The face of the weir should be perpendicular, that is, leaning neither up nor down stream.

4. The crest of the weir should be level so the water passing over it will be of the same depth at all points along the crest.

5. The distance of the crest above the bottom of the pool should be about three times the depth of water flowing over the weir crest, and the sides of the pool should be at a distance from the sides of crest not less than twice the depth of the water passing over the crest.

6. The gage or weir scale may be placed on the upstream face of the weir structure and far enough to one side so that it will be in comparatively still water as shown in Figure 3. The bottom or zero of the scale is at the same height as the weir crest. The scale should be tacked on a 1 inch by 3 inch board a little longer than the scale, and the board then nailed to the structure. This board and the scale can be removed from the structure and reset at any time without in any way damaging the scale. The scales that are made of enamel ware will not stand much bending without cracking the enamel. It has been found that the setting of the scale at one side of the weir as shown gives practically the same results as when it is set in the pool above as is usually directed. It is set with much less trouble, is more permanent and is easier to check.

7. The structure should have the width of the weir crest plainly

marked on the upstream face. The metal parts of the weir should be accurately made and should be carefully placed after the weir box and the weir board are installed.

8· The crest should be placed high enough so the water will fall practically free below the weir. A submergence or back water condition equal to a depth of about 1-16 of the depth of the water over the weir or less has very little effect on the weir discharge and may be neglected in ordinary measurements. Should a close measurement be desired the table of coefficients given may be used. Submerged weirs should be avoided if possible.

9. For accurate measurements the depth over the crest should not be more than one-third the length of the crest. If the depth of water over the weir is greater than one-third of the length of the weir, the discharge as derived from the formula, Q 168.3 LH 3-2 (Where Q is the quantity in miner's inches, L the length of the crest in feet and H the gage height in feet) can be increased by the coefficients as given in the following table:

Table No. I.

Depth in per cent of width	Coefficient
40	1.0075
50	1.018
60	1.032
70	1.047
80	1.063
90	1.080
100	1.096

10. The depth of water over the crest should not be less than about 2 inches as it is difficult to get sufficiently accurate gage readings below this point to give close results. However, a broad crested weir with low gage heights, used where there is little fall will give more reliable results as a rule than can ordinarily be obtained by the use of an orifice using the same amount of head·

Use of the Weir Table.

The depth of water flowing over the weir can be measured in tenths and hundredths of a foot with a weir scale, or it can be measured in inches and fractions of an inch. In the table the head is given both as a decimal of a foot and in inches. To find the discharge of a weir measure the head, or depth of water flowing over it. Look down the column of your table "Head on Crest" till you find the head corresponding to your

measurement, then run across the page to the column headed with the length of your weir crest. This figure will be the discharge of your weir in miner's inches. Example: Suppose that you have a 2½ foot weir and the head is 0.36 of a foot or 4⅜ inches. Look down the column "Head on Crest" to 0.36 then run across the page to column headed 2½ where the discharge is given as 90.2 inches.

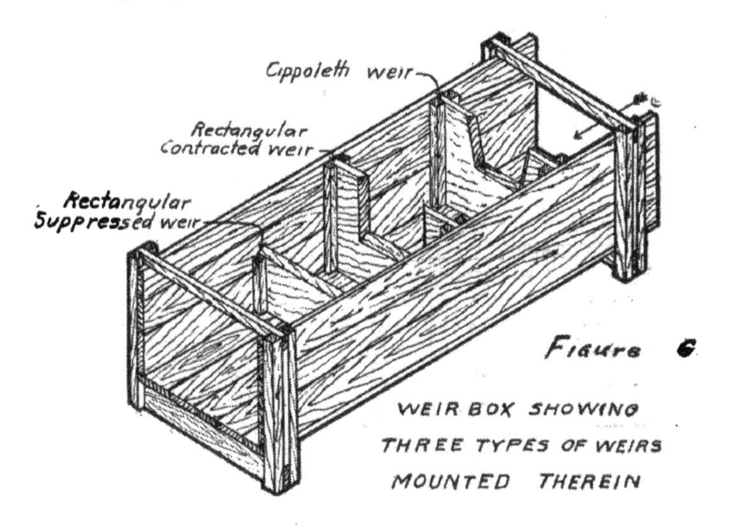

Rectangular Suppressed Weir

If desired, the rectangular suppressed weir can be used in the weir box as shown above, but the Cippoletti weir is usually considered preferable for general work. The table given can be used for a rectangular suppressed weir so that flow over checks and headings may be computed when desirable.

The principal use of the rectangular suppressed weir will be where water is measured over checks and drops. If the measurement is made over the crest with a broad scale according to instructions, the error caused by the velocity of approach is taken up in static head so that the discharge may be computed from a Cippoletti weir table. If the crest of the check or weir in a contracted rectangular weir is long as compared with the depth of the water flowing over it, the error caused by the contraction at either end will be small compared with the total amount when the Cippoletti table is used. In nearly all cases where it is desirable to measure the water passing a check the object is to get an estimate of the water available for use below the check, and for this purpose the use of a check board is sufficiently accurate. If the check board is not level it can be divided into several parts and each part treated as an individual weir.

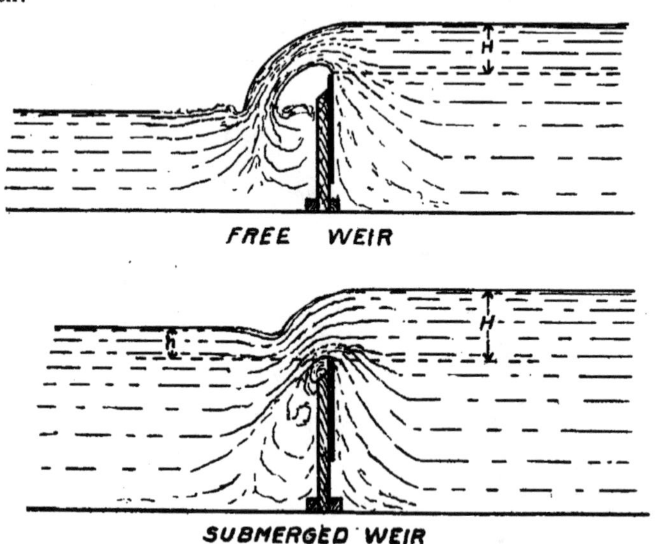

Figure 7 Showing Free and Submerged Weirs

Submerged Weirs

It occasionally happens that, in order to irrigate a piece of high land near the head of a lateral, it is necessary to check the flow of water so that the surface of the water on the lower side of the weir rises above the crest, and the weir becomes a submerged weir; sometimes called a drowned or floating weir. It is evident that under this condition the discharge of the weir cannot be known by reading only the upper gage. In order to get the proper discharge it is necessary to get the elevation of the water surface both above and below the weir with reference to the weir crest.

When the water below the weir is backed up or checked till the surface is higher than the weir crest and the weir begins to submerge, the discharge of the weir is slightly greater than it would be if the flow were free. This excess discharge is gradually diminished as the back water gets higher until the point at which the lower gage reading is about 1-16 of the upper gage. At this point the flow is about the same as if the weir were flowing free. As the water below the weir continues to rise the discharge becomes less and less until the flow finally ceases when the two gage heights become the same. As long as there is a decided difference in the two gage readings the discharge may be obtained with approximate accuracy by the following method:

1. Find the height of the water surface over the weir crest both above and below the weir crest.

2. Find from the weir table what the discharge would be with the weir running free at the gage height as observed on the upper gage.

3. Divide the amount of submergence, or the lower gage reading, by the upper gage reading. This will give a decimal fraction.

4. Select from the coefficient table the value which corresponds to this decimal fraction.

5. Multiply the discharge as shown by the upper gage by this coefficient and the result will be the discharge of the weir as submerged.

6. A submerged weir should never be used where it can possibly be avoided.

Example: On a two foot weir the upper gage reads 0.45 feet. The water surface below the weir stands at 0.26 feet above the weir crest. The discharge of a two foot weir at 0.45 feet is 100.8 miner's inches, (See Weir table); 0.26 divided by 0.45 is equal to 0.577 or approximately 0.58. From Table II, given below, follow down the column headed $\frac{h}{H}$ to 0.50 then run across the table to the column headed 0.08 to complete the 0.58 and the coefficient is found to be 0.780. Multiply the indicated discharge, 100.9, by 0.78 and the actual discharge, 78.6 inches, is obtained.

Table No. II.

TABLE OF COEFFICIENTS FOR SUBMERGED CIPPOLETTI WEIRS

h/H	0.00	0.01	0.02	0.03	0.04	0.05	0.06	0.07	0.08	0.09
0.00	.998	.999	1.000	1.000	1.001	1.001	1.002	1.002	1.001	1.000
0.10	.999	.997	.995	.993	.991	.988	.985	.981	.977	.973
0.20	.969	.965	.961	.957	.953	.948	.943	.939	.935	.931
0.30	.927	.923	.919	.915	.911	.907	.903	.898	.893	.888
0.40	.883	.878	.873	.868	.863	.858	.853	.848	.842	.836
0.50	.830	.824	.818	.812	.806	.800	.794	.787	.780	.773
0.60	.766	.759	.752	.745	.738	.730	.722	.714	.706	.697
0.70	.688	.679	.670	.660	.650	.640	.629	.618	.607	.595
0.80	.583	.571	.558	.545	.532	.518	.503	.487	.471	.453
0.90	.435	.415	.395	.372	.347	.319	.287	.253	.205	.143

* H=The gage height above the weir crest.
h=The gage below the weir crest.

Another method of getting the discharge of a submerged weir is by the use of the diagram shown in Figure 8. In order to use this diagram the upper and lower gage height must be known, the same as in using the table, then proceed as shown in the following example:

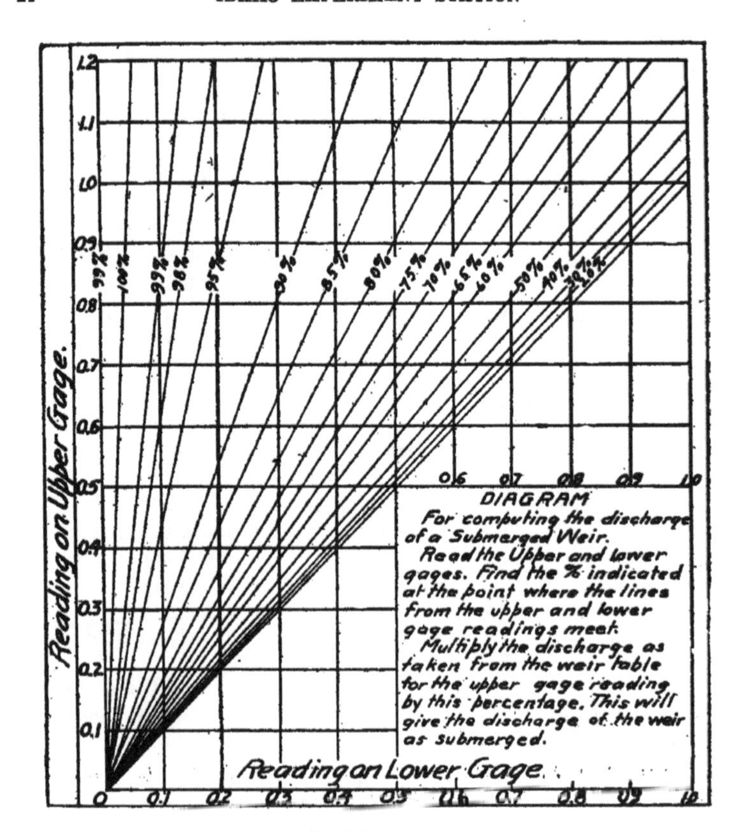

FIG. 8. SUBMERGED WEIR DIAGRAM.

Suppose that the gage above a two foot weir reads 0.70 feet and the surface below the weir is 0.40 feet above the crest. Referring to the diagram and noting where the line for 0.70 for the upper gage reading crosses the line for 0.40 the lower gage reading, it will be seen that they cross near the diagonal line marked 80 per cent or at about 78 per cent. Taking the discharge of a 2 foot free weir with a head of 0.70 feet, we have 197.6 inches from the weir table, and multiplying this by 78 per cent we get the true discharge of 154.1 inches for the weir that has a gage height of 0.70 feet and is submerged 0.40 feet.

Care should be taken that there be enough difference in the readings of the upper and the lower surfaces to give a location on the diagram where the values do not change too rapidly, for instance, 0.80 feet on the upper and 0.77 feet on the lower would give a location on the diagram where it would be difficult to give the proper percentage for a multiplier. Under such cases the checks might be changed so as to lower the water surface below the weir, thus making a greater difference in the two gage readings while still leaving the weir submerged.

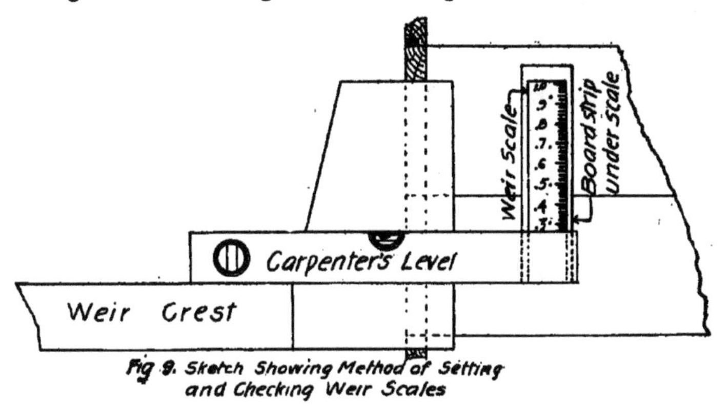

Fig 9. Sketch Showing Method of Setting and Checking Weir Scales

Figure 12 shows the method of measuring the backwater on a submerged weir. The scale, held in a vertical position, is pressed against the downstream face of the weir plate in such a way that the corner of the scale nearest the center of the ditch rests on the end of the weir crest, The backwater then is easily read as shown in the figure mentioned. If the weir crest is not level then it may be taken in two sections and a measurement can be made at each end.

The submergence of a weir is usually caused by the canal below becoming filled more or less with grass, moss or weeds. When possible this condition should not be permitted. The use of the submerged weir for the measurement of water is not recommended but it gives fairly accurate results and if the measurements of the depth of the water above and below the crest are carefully made the resulting calculations are far better than are usually obtained from a rider's estimate.

Submerged Weir Diagram

The diagram shown in Figure 8 was derived from experiments conducted by Francis, Ftele, Stearns and Stevens and its use is considered reliable if the difference in the gage heights is not so small as to make the percentage used less than 50. The use of this diagram enables the rider to make a quick and close computation of the discharge over a submerged weir.

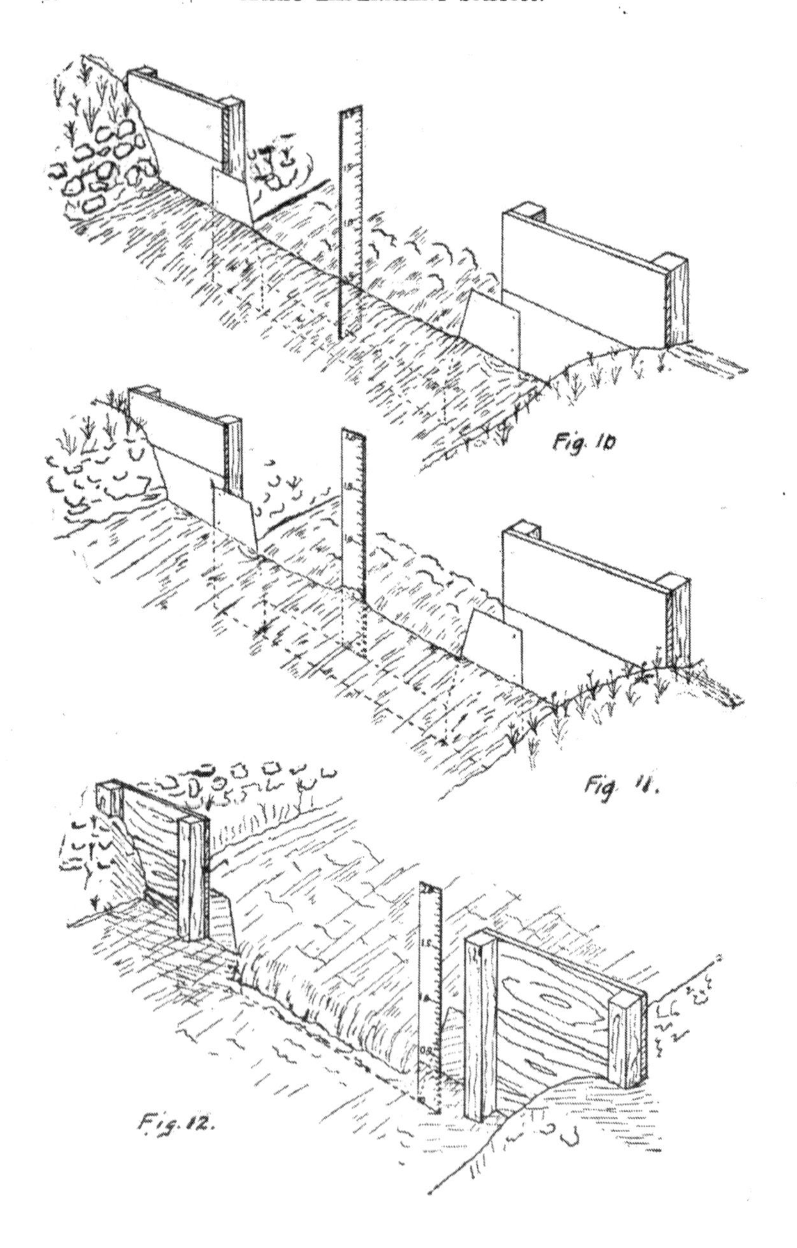

Fig. 10

Fig. 11.

Fig. 12.

Checking Weir Gages

The position of the gage should be checked frequently to determine if the weir box has settled out of level, or if the gage or other parts of the weir have been changed. This may easily be done by the use of a carpenter's level (See Figure 9) or it may be closely approximated by turning the water out of the lateral and noting at what point on the gage the water stands when the point of zero flow takes place on the weir. Care must be taken in doing this as the water will not flow over the crest till it is about 0.015 of a foot above the actual crest level. This method will also show whether or not the weir crest is level.

Velocity of Approach

Where the weir pool is made too narrow and too shallow, or where the pool has become filled with silt, the water as it flows towards the weir has considerable velocity. This is known as velocity of approach. This velocity makes the weir discharge more water with the same gage height than it would if the water were coming from a still pool. It is for the correction of this that the weir pool is made wide and deep. Where this velocity occurs, a correction can be made for it in the following manner. Using a thin weir scale about two and one-half or three inches wide; place the scale on the weir crest with the sharp edge cutting the current as shown in Figure 10, and then turn it parallel with the edge of the weir crest letting the water climb up the scale as shown in Figure 11, note the reading of the water on the scale. This reading is the one to use in computing the discharge from the weir table. Do not hold the scale in this position longer than is necessary to make the reading because the width of the scale blocks the water and will in a short time raise the water in the pool and increase the reading. If the reading is not obtained the first time, try it until a satisfactory reading is secured. It has been found that this method gives the same reading as if the weir pool were cleaned and a reading were made under normal conditions. If the weir is normal and has no scale mounted thereon reading can be taken in this manner at any time. It requires a broad scale and it must be held perpendicular over the weir crest. An ordinary carpenter's rule is too narrow to be used for reading depths accurately. The theory of this method is that the scale being placed in the water with the broad face against the stream, checks the water and thus takes up the velocity of approach and so changes the velocity head to static head, which is read on the gage. In case the velocity is not evenly distributed across the weir, the crest can be divided into two or more sections and a separate measurement can be made of each and the results added. It sometimes happens that a weir crest has settled so that one side is lower than the other. In this case it is evident that a reading made in the center of the weir would not give the proper results. To get the correct measurement, divide the crest into two or more sections and sum the results· Do not take the mean gage height of the sections as that would not give the proper discharge. A little experience will enable one to use this method to good advantage. It is not intended that this method of measurement is to take the place of well constructed weirs and pools, but is to be used when the pool is out of order or when there is no gage on the structure by which the proper gage height can be obtained.

Figure 13—Using a scale in taking the depth of water over a weir.

In Figure 13 is shown the method of taking the depth of water over a weir using a scale. In this case the gage on the weir structure did not show the proper reading because it was situated off in one corner where it did not get the effect of the velocity of approach, of which there was considreable. The water from the heading above ran at an angle across the pool and the highest velocity passed over the center of the weir. This would necessitate that there be at least two or perhaps more readings made on this weir in order to get the proper discharge. The gage installed on the heading read 0.19 feet and the scale reading over the center of the weir read 0.22 feet. This pool was then deepened by raising the weir board 0.65 feet and readings were again taken, the discharge of the heading not being disturbed· The two gages then read the same. They were both 0.21 feet. Figure 14 shows these readings being taken. After these readings were taken the head of water was increased and the readings as shown in Figure 15 were taken. They were 0.70 feet for the gage on the structure and 0.69 feet for the gage as seld over the crest of the weir, thus showing how close the two will tally when the weir is in proper

Figure 14—Taking the reading after raising the weir board.

Figure 15—Taking readings after increasing the head of water.

condition. Figure 16 shows a weir having too high a velocity of approach, which renders it valueless as a measuring device. This easily could be corrected by raising the weir crest until the velocity of approach was reduced to less than one-half foot per second.

These illustrations should give a good idea of the proper use of a scale by a rider for keeping close check on the flow of water in the canals along his ride.

Figure 16—The velocity of approach too high.

The Submerged Orifice

The structure for the submerged orifice is built the same as for a weir, but instead of placing a weir crest in the front heading an opening known as an orifice is placed therein as shown in Figure 17.

This opening may be of any required dimensions and of any shape, tho for convenience of computation certain standard dimensions are usually selected. (See Table V.) The orifice is not used as generally as the weir. This is due to certain inherent disadvantages. First, the orifice structure is such that it gathers trash which tends to check the flow and hence to destroy the accuracy of the measurement. Second, there is a chance for inaccuracy on low heads, that is, where there is but little difference between the upper and lower gage readings because the relative discharge for this small difference is so great that a slight error in reading the gage makes a very great difference in the result of the computed discharge. Third, unless some special provision is made the submerged orifice is not adapted to passing large quantities of water; it will pond the water above the orifice so that damage from overflow is liable to be done to the canal or the heading. In the case of the weir the proportional discharge is increased as the head increases and the excess flow is automatically taken care of by passing over the weir. This is graphically shown in Figure 18, where the discharge curve of a two foot weir is plotted on the same scale as the curve for an orifice with a one foot opening. It will be noted that for a gage height of 0.10 feet the discharge of the orifice is about 8 times as great as the discharge of the weir, this proportional discharge is gradually reduced till at the gage height of 0.75

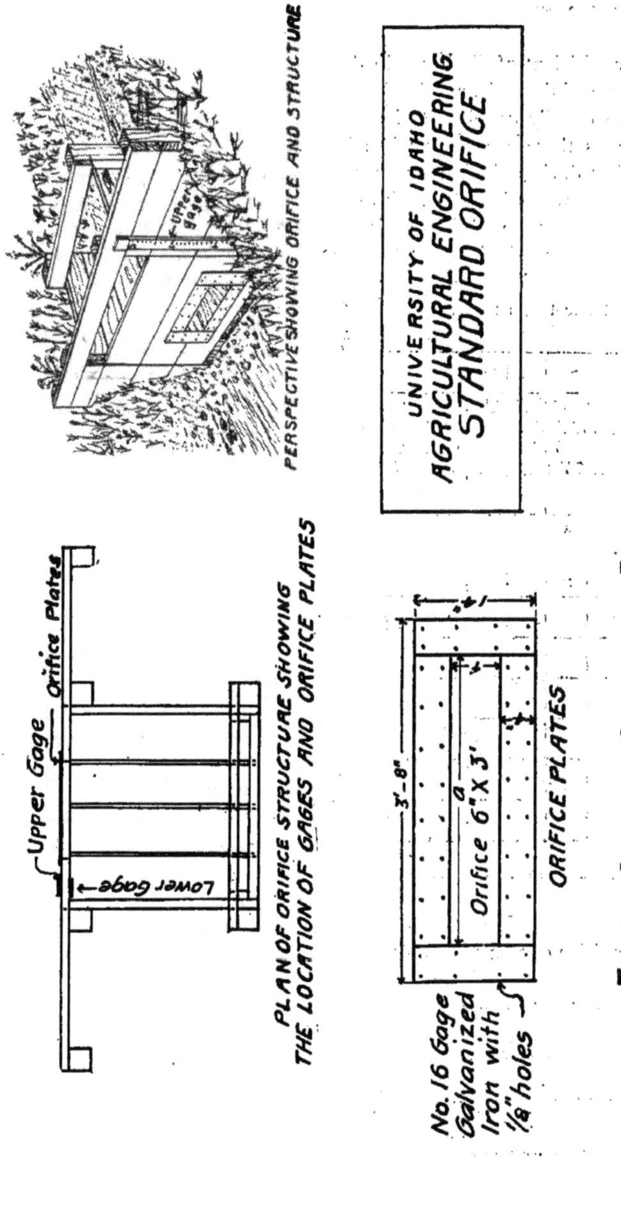

Upper Gage

Orifice Plates

Lower Gage

PLAN OF ORIFICE STRUCTURE SHOWING
THE LOCATION OF GAGES AND ORIFICE PLATES

Upper gage

PERSPECTIVE SHOWING ORIFICE AND STRUCTURE

3'-8"

No. 16 Gage
Galvanized
Iron with
1/8" holes

a

Orifice 6" X 3'

1'-4"

ORIFICE PLATES

UNIVERSITY OF IDAHO
AGRICULTURAL ENGINEERING
STANDARD ORIFICE

FIG. 17 SHOWING STANDARD ORIFICE AND ORIFICE PLATES

feet they are equal, and at one foot the discharge of the weir is 35 per cent greater than the discharge of the orifice.

The coefficient of discharge in the orifice is much more uncertain than in the case of the weir, and is affected by a greater variety of factors that are not so easily regulated· Notwithstanding the above mentioned disadvantages there are times when it is desirable to use the orifice as a measuring device. This may occur where it is imperative to save head, or hold the water level as high as possible in the canal. In this case it may be necessary to sacrifice accuracy for the sake of saving head. There are times when it is desirable to combine a canal heading with a measuring device in which case an orifice can well be used because the heading shuts out the trash and regulates the flow.

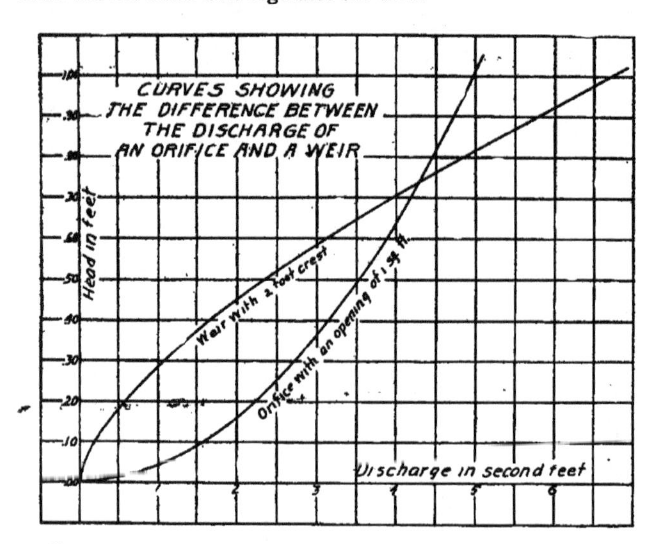

FIG. 18 CURVES TO GIVE A COMPARISON OF ORIFICE AND WEIR DISCHARGE.

Rules Governing the Use of the Orifice

The orifice opening should be regular in shape, and should have sharp edges. The pressure head should be not less than 0.10 of a foot.

The depth of submergence of the orifice should not be less than the height of the orifice, and a submergence of twice the height of the orifice is preferable·

There must be two gages, one of which should be set on the head-wall above and to one side of the orifice, and the other on the head-wall to one side and below the orifice. These gages should be set with their zero marks at the same elevation. This may be at any desired point so it will always be covered with water when the orifice is in use.

Where a canal gate or heading is used for an orifice to measure water the coefficient of discharge must be determined for each different condition either by measuring the water over a weir or by a current meter measurement, if any degree of accuracy is required. This is because the discharge coefficient changes with the form and kind of orifice and in many cases with the depth of water and the water pressure. For this reason if good measurements are desired the standard orifice structure as shown should be used, and the discharges may then be taken from the table as given.

Orifice Table

In using the table find the difference between the reading of the upper and the lower gages. This will be the "effective head" as used in the table. In the succeeding columns will be found the discharges for orifices having openings of different sizes with effective heads as shown by the gages. Discharge for orifices of larger sizes may be found by multiplying the discharge for a one foot orifice by the size of the orifice desired.

Example: Suppose that the upper gage reads 1.25 feet and the lower gage reads 0.83 feet, then the effective gage reading would be 0.42 feet. If the orifice opening is 3 inches wide and 12 inches long, then the orifice would have an area of ¼ square foot. Referring to the table opposite 0.43 foot and in calumn headed ¼ square foot the discharge is given as 40.8 miner's inches. The coefficient used in this table is 0.62. If any other coefficient is desired, divide the given discharge by 0.62 and multiply the result by the desired coefficient and the required discharge is obtained.

For the convenience of those accustomed to measuring the heads in inches the "Head" is given in feet and in inches in both the weir and orifice tables.

Float Measurements

It often happens that it may be neccessary to know how much water is flowing in a canal or lateral and that there is no measuring device installed. In that case a fairly accurate measurement can be made by the use of floats. To do this, select a part of the stream where the flow is even and straight and the width and depth are uniform. This is necessary because curves and holes create cross currents in the stream. The length of the section to be measured will depend on conditions but it should be at least 4 or 5 times the distance across the stream. Select and mark two points along the bank that are at suitable distance from each other. Opposite each point lay a pole or stretch a string or rope across the stream at right angles to the direction of flow. By means of marks on the rod or string divide the distance across the stream into any number of sections. These sections may be even in distance or not to suit the conditions. Measure the depth of water under the center of each section. (See Figure 19).

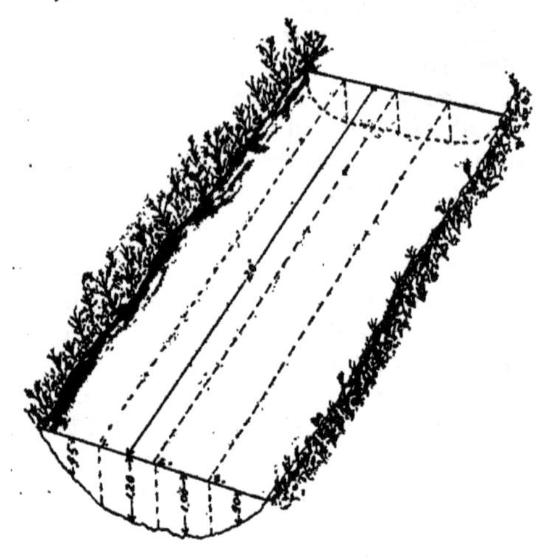

Fig. 19. Showing Method of Dividing Stream Into Sections For Float Measurement.

A set of notes somewhat as follows should be kept:

Table III.

Form of Notes for Float Measurements

Section	Width of Section	Depth at Center	Area	Length of Section	Time of Float	Velocity of Water	Discharge
Ft.	Ft.	Ft.	Sq. Ft.	Ft.	Seconds	Ft. per Sec.	Sec. Ft.
0-1	1	.45	.45	20	16	1.25	.55
1-2	1	1.20	1.20	20	10	2.00	2.40
2-3	1	1.00	1.00	20	8	2.50	2.50
3-4	1	.40	.40	20	15	1.33	.53
						Total	.........5.98

Select a material for a float that will be almost entirely submerged as a green Russian thistle or a bottle or can partially filled with water. This is to avoid its being materially affected by the air currents and so it will acquire as near the mean velocity of the water as possible. Place this float in the stream above the upper section and opposite one of the divisions, and let it float down between the two marked sections and note the number of seconds required to pass between the two lines. A stop watch is best for this work but with a little practice an ordinary watch will enable one to get good results. Should the float tend to move out of the desired line of flow a light pole may be used to bring it into place, but great care must be taken not to change its velocity in this operation. One or more tests should be made on each section. The notes are completed as follows:

The width of a cross section multipiled by the measured depth gives the area in square feet. The length of the section divided by the time in seconds it took the float to pass between the lines will give the velocity per second in feet. The area of the section in square feet multiplied by the velocity per second will give the discharge of that section in cubic feet per second or in second feet. Adding the discharges of all the sections will give the apparent discharge of the stream in second feet. This multiplied by the proper coefficient will give the true discharge in second feet. To reduce second feet to miner's inches multiply by 50.

The apparent discharge as derived by this method must be multiplied by a coefficient to get the true discharge. This is because the bottom of the stream moves slower than the top and the float does not get the mean velocity of the entire flow of the water. The coefficient to be used in any case will depend on the condition of the stream bed and the character of the float used. It will vary from 0.80 to 0.90. If the stream bed is rough and the float is carried at the surface of the water the coefficient will be small, but on the other hand, if the bed is smooth and the float reaches close to the bottom and still flows free, the coefficient will be high. If in the example given, we take the value of 0.85 then the true discharge will be; 5.98 second feet multiplied by 0.85 which equals 5.08 second feet. This is multiplied by 50 to reduce to miner's inches.

As the coefficient to be used with a float measurement cannot be definitely known, the float measurements are apt to be somewhat in error. This error is not serious as compared with an estimate made simply by observation.

Table IV.

Discharge Data, Dimensions and Lumber for Cippoletti Weir Box

Cippoletti Weir			Structures				
Crest length	Maximum depth of water on crest	Maximum discharge	Headwall height b	Side height c	Length d	Floor width e	Approximate quantity of lumber board measure
Ft.	Ft.	Sec. Ft.	Ft.	Ft.	Ft.		Ft.
1.0	0.33	0.64	3.5	3.0	4.0	2.0	130
1.5	0.50	1.79	3.5	3.0	4.0	2.5	135
2.0	0.67	3.69	3.5	3.0	4.0	3.0	145
3.0	1.00	10.00	4.0	4.0	4.0	4.0	250

Taken from U. S. Reclamation Service designs.

Table V.

Dimensions and Lumber for Standard Sizes of Submerged Rectangular Orifices*

Size of Orifice							
Height f	Length a	Area a x f	Headwall height b	Side height c	Structure length d	Floor width e	Approximate quantity of lumber board measure
Ft.	Ft.	Sq. Ft.	Ft.	Ft.	Ft.	Ft.	Ft..
0.25	1.00	0.25	3.0	2.5	4.0	2.0	150
	2.00	0.50	3.0	2.5	4.0	3.0	170
0.50	1.00	0.50	3.0	2.5	4.0	2.0	150
	2.00	1.00	3.0	2.5	4.0	3.0	170
	3.00	1.50	3.5	2.5	4.0	4.0	210
0.75	1.33	1.00	3.0	2.5	4.0	2.0	150
	2.00	1.50	3.0	2.5	4.0	3.0	170
	2.67	2.00	3.5	3.0	4.0	4.0	200

*Taken from U. S. Reclamation Service.

Hydraulic Equivalents

1 cubic foot = approximately 7½ gallons.

1 cubic foot of water weights 62.4 pounds.

1 cubic foot per second = 448.8 gallons per minute
= 1.9835 acre feet per day
= 50 miner's inches (Idaho statutes).

1 second foot falling 8.81 ft. = 1 horse power
= 33000 foot pounds per minute.

1 second foot falling 11. ft. = 1 horse power, 80% efficiency.

1 second foot for 1 year will cover 1 square mile 1.131 feet deep.

1 acre foot = the volume of water that will cover an acre 1 foot deep.

1 acre inch = the volume of water that will cover an acre 1 inch deep.

Those who wish to do current meter work should send to the U. S. Geological Survey, Washington, D. C., for Water Supply Paper No. 94 entitled "Hydrographic Manual, in which detailed information is given on the use of the current meter.

To Water Masters and Riders

Measure and keep records of all deliveries.

Ask no favors of the water users and show none in your measurements or deliveries to any one.

Be as accurate as possible in all deliveries.

Keep the headings free from moss, weeds and trash.

Keep as near a steady flow at the lower end of the laterals as possible. If the last user is well served you will have solved your hardest problem.

Learn the amount of water that is due each user and see that he gets it.

By proper administration and the adjustment of differences you may become a community peacemaker. It costs more for a law suit than for a good ditch rider and the suit often gets nowhere.

Be polite and considerate to all users· Do not lose your temper; the greater the provocation the more need for restraint.

Always carry a ride's book and a weir scale when on duty. You can the better answer objections and correct mistakes.

Answer with a smile.

To Water Users

Remember that the water master and the rider are human.

Do not expect them to serve you better than they can your neighbor.

Excess water delivered to you deprives some other user of his rightful share.

If the water master or the rider is not giving good service get another, but pay for good service as you would expect to be paid· Do not expect good work for poor pay.

Your water is your most valuable asset. Do not trust it's management to an inexperienced man and then complain of poor service.

Do not quarrel or complain but inform yourself and explain; it may be that the other fellow is right sometimes.

Keep good will in the community; it solves many troubles.

Do not tamper with headgates; it is dishonest and criminal. The rider is supposed to be doing his duty and will attend to the change when needed.

Remember that the ditch rider is acting under instructions from his superiors·

A good rider is your friend; treat him as such.

Acknowledgements

In preparing this bulletin, use has been made of illustrations and weir table furnished by the U. S. Reclamation Service. The submerged weir diagram was adapted from a report by J. C. Stevens of Portland, Oregon.

A Brief List of Publications on Water Measurement

1. River Discharge, by Hoyt & Grover, Wiley & Sons, New York.
2. Hydrographic Manual of the U. S. Geological Survey, Water Supply Paper No. 94, Washington, D. C.
3. Accuracy of Stream Measurements, No. 95 Water Supply Papers, U. S. Geological Survey, Washington, D. C.
4. Water Measurements, Montana Agricultural College, Bozeman, Mont.
5. How to Measure Water, State College of Washington, Pullman, Wn.
6. Practical Information on the Measurement of Irrigation Water, Utah Agricultural College Experiment Station, Logan, Utah.
7. The Measurement and Division of Water, Colorado Agricultural College, Fort Collins, Colorado.
8. Flow of Water in Irrigation Channels, (U. S. D. A. Department Bulletin 194), Price 25c.

Reference Books on Irrigation Practice and Engineering

1. Operation and Maintenance of Irrigation Systems by S. T. Harding, McGraw Hill Book Co., N. Y.
2. Principles of Irrigation Practice by John A. Widtsoe, McMillan & Co., N. Y.
3. How to Build Small Irrigation Ditches, U. S. Department of Agriculture (Farmer's Bulletin 158).
4. Practical Information for Beginners in Irrigation, U. S. Department of Agriculture (Farmer's Bulletin).
5. Irrigation Practice and Engineering, 3 Volumes by B. A. Etcheverry, McGraw Hill Book Co., N. Y.
6. Irrigation and Drainage by F. H. King, The McMillan Co., N. Y.
7. Use of Water in Irrigation by Samuel Fortier, McGraw-Hill Book Co., N. Y.

Table VI.

DISCHARGE OVER CIPPOLETTI WEIRS

Quantities in Miner's Inches

Head on Crest in Ins.	Head on Crest in Feet	Width of Weir Crest in Feet												
		1′	1½′	2′	2½′	3′	3½′	4′	5′	6′	7′	8′	9′	10′
⅛	.01	0.2	0.3	0.3	0.4	0.5	0.6	0.7	0.8	1.0	1.2	1.4	1.5	1.7
¼	.02	0.5	0.7	1.0	1.2	1.4	1.7	1.9	2.4	2.9	3.4	3.8	4.3	4.8
⅜	.03	0.9	1.3	1.8	2.2	2.7	3.1	3.6	4.4	5.3	6.2	7.1	8.0	8.9
½	.04	1.4	2.0	2.7	3.5	4.1	4.8	5.4	6.8	8.2	9.5	10.9	12.2	13.6
⅝	.05	1.9	2.8	3.8	4.8	5.7	6.6	7.6	9.6	11.4	13.3	15.2	17.1	19.0
¾	.06	2.5	3.8	5.0	6.2	7.5	8.8	10.0	12.5	15.0	17.5	20.0	22.5	25.0
⅞	.07	3.1	4.7	6.3	7.8	9.4	10.9	12.6	15.7	18.8	22.0	25.1	28.3	31.4
1	.08	3.8	5.7	7.6	9.5	11.4	13.3	15.2	19.0	22.8	26.6	30.4	34.2	38.0
⅛	.09	4.6	6.8	9.1	11.5	13.7	16.0	18.3	22.8	27.4	32.0	36.6	41.1	45.7
¼	.10	5.4	8.0	10.7	13.4	16.0	18.7	21.4	26.8	32.1	37.4	42.8	48.2	53.5
⅜	.11	6.2	9.2	12.3	15.5	18.5	21.7	24.6	30.8	37.0	43.1	49.3	55.4	61.6
½	.12	7.0	10.5	14.0	17.5	21.1	24.5	28.1	35.1	42.1	49.1	56.2	63.2	70.2
9/16	.13	7.9	11.8	15.8	19.8	23.7	27.6	31.6	39.5	47.4	55.3	63.2	71.1	79.0
⅝	.14	8.8	13.2	17.7	22.0	26.5	30.8	35.3	44.2	53.0	61.8	70.6	79.5	88.3
¾	.15	9.8	14.7	19.6	24.4	29.3	34.2	39.1	48.9	58.7	68.5	78.2	88.0	97.8
1⅞	.16	10.7	16.2	21.5	26.9	32.3	37.7	43.1	53.8	64.6	75.4	86.2	96.9	107.7
2	.17	11.8	17.7	23.6	29.5	35.4	41.3	47.2	59.0	70.7	82.5	94.3	106.1	117.9
⅛	.18	12.8	19.2	25.7	32.0	36.5	44.8	51.3	64.2	77.0	89.8	102.6	115.5	128.3
¼	.19	13.9	20.9	27.8	34.8	41.8	48.7	55.7	69.6	83.5	97.4	111.4	125.3	139.2
⅜	.20	15.0	22.5	30.1	37.5	45.1	52.5	60.1	75.2	90.2	105.2	120.2	135.3	150.3
		1′	1½′	2′	2½′	3′	3½′	4′	5′	6′	7′	8′	9′	10′

		1′	1½′	2′	2½′	3′	3½′	4′	5′	6′	7′	8′	9′	10′
½	.21	16.2	24.3	32.3	40.5	48.5	56.7	64.6	80.8	97.0	113.1	129.3	145.4	161.6
⅝	.22	17.3	26.0	34.6	43.3	51.9	60.6	69.2	86.6	103.9	121.2	138.5	155.8	173.1
¾	.23	18.5	27.8	37.0	46.2	55.4	64.7	73.9	92.4	110.9	129.4	147.8	166.3	184.8
⅞	.24	19.7	29.6	39.4	49.3	59.1	69.0	78.8	98.6	118.3	138.0	157.7	177.4	197.1
3	.25	20.9	31.4	41.9	52.3	62.8	73.2	83.8	104.8	125.7	146.6	167.6	188.6	209.5
⅛	.26	22.2	33.3	44.4	55.5	66.6	77.7	88.8	110.0	132.2	155.4	177.6	199.8	222.0
¼	.27	23.5	35.2	47.0	58.8	70.5	82.2	94.0	117.5	141.0	164.5	188.0	211.5	235.0
⅜	.28	24.8	37.2	49.6	62.0	74.4	86.8	99.2	124.0	148.8	173.6	198.4	223.2	248.0
½	.29	26.1	39.2	52.2	65.2	78.3	91.3	104.4	130.5	156.6	182.7	208.8	234.9	261.0
⅝	.30	27.5	41.2	55.0	68.8	82.5	96.3	110.0	137.5	165.0	192.5	220.0	247.5	275.0
¾	.31	28.8	43.2	57.6	72.0	86.4	100.8	115.2	144.0	172.8	201.6	230.4	259.2	288.0
⅞	.32	30.2	45.3	60.4	75.5	90.6	105.7	120.8	151.0	181.2	211.4	241.6	271.8	302.0
4	.33	31.7	47.6	63.4	79.2	95.1	111.0	126.8	158.5	190.2	221.9	253.6	285.3	317.0
⅛	.34	33.1	49.6	66.2	82.8	99.3	115.8	132.4	165.5	198.6	231.7	264.8	297.9	331.0
¼	.35	34.6	51.9	69.2	86.5	103.8	121.1	138.4	173.0	207.6	242.2	276.8	311.4	346.0
⅜	.36	36.1	54.2	72.2	90.2	108.3	126.4	144.4	180.5	216.6	252.7	288.8	324.9	361.0
½	.37	37.6	56.4	75.2	94.0	112.8	131.6	150.4	188.0	225.6	263.2	300.8	338.4	376.0
9/16	.38	39.1	58.6	78.2	97.8	117.3	136.8	156.4	195.5	234.6	273.7	312.8	351.9	391.0
⅝	.39	40.6	60.9	81.2	101.5	121.8	142.1	162.4	203.0	243.6	284.2	324.8	365.4	408.0
¾	.40	42.2	63.3	84.4	105.5	126.6	147.7	168.8	211.0	253.2	295.4	337.6	379.8	422.0
⅞	.41	43.8	65.7	87.6	109.5	131.4	153.3	175.2	219.0	262.8	306.6	350.4	394.2	438.0
5	.42	45.4	68.1	90.8	113.5	136.2	158.9	181.6	227.0	272.4	317.8	363.2	408.6	454.0
⅛	.43	47.0	70.5	94.0	117.5	141.0	164.5	188.0	235.0	282.0	329.0	376.0	423.0	470.0
¼	.44	48.6	72.9	97.2	121.5	145.8	170.1	194.4	243.0	291.6	340.2	388.8	437.4	486.0
⅜	.45	50.4	75.6	100.8	126.0	151.2	176.4	201.6	252.0	302.4	352.8	403.2	453.6	504.0
		1′	1½′	2′	2½′	3′	3½′	4′	5′	6′	7′	8′	9′	10′

Table No. VI. (Continued)

DISCHARGE OVER CIPPOLETTI WEIRS

Quantities in Miner's Inches

Head on Crest in Ins.	in Feet	Width of Weir Crest in Feet												
		1′	1½′	2′	2½′	3′	3½′	4′	5′	6′	7′	8′	9′	10′
5½	.46	52.0	78.0	104.0	130.0	156.0	182.0	208.0	260.0	312.0	364.0	416.0	468.0	520.0
⅝	.47	53.8	80.7	107.6	134.5	161.4	188.3	215.2	269.0	322.8	376.6	430.4	484.2	538.0
¾	.48	55.6	83.4	111.2	139.0	166.8	194.6	222.4	278.0	333.6	389.2	444.8	500.4	556.0
⅞	.49	57.3	86.0	114.6	143.2	171.9	200.6	229.2	286.5	343.8	401.1	458.4	515.7	573.0
6	.50	59.1	88.6	118.2	147.8	177.3	206.8	236.4	295.5	354.6	413.7	472.8	531.9	591.0
⅛	.51	60.9	91.4	121.8	152.2	182.7	213.2	243.6	304.5	365.4	426.3	487.2	548.1	609.0
¼	.52	62.8	94.2	125.6	157.0	188.4	219.8	251.2	314.0	376.8	439.6	502.4	565.2	628.0
⅜	.53	64.6	96.9	129.2	161.5	193.8	226.1	258.4	323.0	387.6	452.2	516.8	581.4	646.0
½	.54	66.5	99.8	133.0	166.2	199.5	232.8	266.0	332.5	399.0	465.5	532.0	598.5	665.0
⅝	.55	68.4	102.6	136.8	171.0	205.2	239.4	273.6	342.0	410.4	478.8	547.2	615.6	684.0
¾	.56	70.2	105.3	140.4	175.5	210.6	245.7	280.8	351.0	421.2	491.4	561.6	631.8	702.0
⅞	.57	72.2	108.3	144.4	180.5	216.6	252.7	288.8	361.0	433.2	505.4	577.6	649.8	722.0
7	.58	74.2	111.3	148.4	185.5	222.6	259.7	296.8	371.0	445.2	519.4	593.6	667.8	742.0
⅛	.59	76.1	114.2	152.2	190.2	228.3	266.4	304.4	380.5	456.6	532.7	608.8	684.9	761.0
¼	.60	78.1	117.2	156.2	195.2	234.3	273.4	312.4	390.5	468.6	546.7	624.8	702.9	781.0
⅜	.61	80.1	120.2	160.2	200.2	240.3	280.4	320.4	400.5	480.6	560.7	640.8	720.9	801.0
½	.62	82.1	123.2	164.2	205.2	246.3	287.4	328.4	410.5	492.6	574.7	656.8	738.9	821.0
9/16	.63	84.2	126.3	168.4	210.5	252.6	294.7	336.8	421.0	505.2	589.4	673.6	757.8	842.0
⅝	.64	86.2	129.3	172.4	215.5	258.6	301.7	344.8	431.0	517.2	603.4	689.6	775.8	862.0
¾	.65	88.2	132.3	176.4	220.5	264.6	308.7	352.8	441.0	529.2	617.4	705.6	793.8	882.0
⅞	.66	90.3	135.4	180.6	225.7	270.9	316.0	361.2	451.5	541.8	632.1	722.4	812.7	903.0
8	.67	92.4	138.6	184.8	231.0	277.2	323.4	369.6	462.0	554.4	646.8	739.2	831.6	924.0
⅛	.68	94.5	141.8	189.0	236.2	283.5	330.8	378.0	472.5	567.0	661.5	756.0	850.5	945.0
¼	.69	96.7	145.0	193.4	241.8	290.1	338.4	386.8	483.5	580.2	676.9	773.6	870.3	967.0
⅜	.70	98.8	148.2	197.6	247.0	296.4	345.8	395.2	494.0	592.8	691.6	790.4	889.2	988.0
		1′	1½′	2′	2½′	3′	3½′	4′	5′	6′	7′	8′	9′	10′

		1′	1½′	2′	2½′	3′	3½′	4′	5′	6′	7′	8′	9′	10′
½	.71	101.0	151.5	202.0	252.5	303.0	353.5	404.0	505.0	606.0	707.0	808.0	909.0	1010.0
⅝	.72	103.2	154.8	206.4	258.0	309.6	361.2	412.8	516.0	619.2	722.4	825.6	928.8	1032.0
¾	.73	105.3	158.0	210.6	263.2	315.9	368.6	421.2	526.5	631.8	737.1	842.4	947.7	1053.0
⅞	.74	107.6	161.4	215.2	269.0	322.8	376.6	430.4	538.0	645.6	753.2	860.8	968.4	1076.0
9	.75	109.8	164.7	219.6	274.5	329.4	384.3	439.2	549.0	658.8	768.6	878.4	988.2	1098.0
⅛	.76	112.1	168.2	224.2	280.2	336.3	392.4	448.4	560.5	672.6	784.7	896.8	1008.0	1121.0
¼	.77	114.3	171.4	228.6	285.8	342.9	400.0	457.2	571.5	685.8	800.1	914.4	1028.7	1143.0
⅜	.78	116.6	174.9	233.2	291.5	349.8	408.1	406.4	583.0	600.0	816.2	932.8	1049.4	1166.0
½	.79	118.8	178.2	239.6	297.0	356.4	415.8	475.2	594.0	712.8	831.6	950.4	1069.2	1188.0
⅝	.80	121.2	181.8	242.4	303.0	363.6	424.2	484.8	606.0	727.2	848.4	969.6	1090.8	1212.0
10¼	.85	132.9	199.4	265.8	332.2	398.7	465.2	531.6	664.5	797.4	930.3	1063.2	1196.1	1329.0
¾	.90	145.0	217.5	290.0	362.5	435.0	507.5	580.0	725.0	870.0	1015.0	1160.0	1305.0	1450.0
11⅜	.95	157.5	236.2	315.0	393.8	472.5	551.2	630.0	787.5	945.0	1102.5	1260.0	1417.5	1575.0
12	1.00	170.3	255.4	340.6	425.8	510.9	596.0	681.0	851.5	1021.8	1192.1	1362.4	1532.7	1703.0
⅝	1.05	183.5	275.2	367.0	458.8	550.5	642.0	734.0	917.5	1101.0	1284.5	1468.0	1651.5	1835.0
13¼	1.10	197.0	295.5	394.0	492.5	591.0	689.5	788.0	985.0	1182.0	1379.0	1576.0	1773.0	1970.0
¾	1.15	210.8	316.2	421.6	527.0	632.4	737.8	843.2	1054.0	1264.8	1475.6	1686.4	1897.2	2108.0
14⅜	1.20	225.0	337.5	450.0	562.5	675.0	787.5	900.0	1125.0	1350.0	1575.0	1800.0	2025.0	2250.0
15	1.25	239.4	359.1	478.8	598.5	718.2	837.9	957.6	1197.0	1436.4	1675.8	1915.2	2154.6	2394.0
⅝	1.30	254.1	381.2	508.2	635.2	762.3	889.4	1016.4	1270.5	1524.6	1778.7	2032.8	2286.9	2541.0
16¼	1.35	269.2	403.8	538.4	673.0	807.6	942.2	1076.8	1346.0	1615.2	1884.4	2153.6	2422.8	2692.0
¾	1.40	284.7	427.0	569.4	711.8	854.1	996.4	1138.8	1423.5	1708.2	1992.9	2277.6	2562.3	2847.0
17⅜	1.45	300.3	450.4	600.6	750.8	909.0	1051.0	1201.2	1501.5	1801.8	2102.1	2402.4	2702.7	3003.0
18	1.50	316.2	474.3	632.4	790.5	948.6	1106.7	1264.8	1581.0	1897.2	2213.4	2529.6	2845.8	3162.0
		1′	1½′	2′	2½′	3′	3½′	4′	5′	6′	7′	8′	9′	10′

Table No. VII

DISCHARGE TABLE FOR STANDARD SUBMERGED ORIFICE

Discharge $=AC\sqrt{2gh}$; A=Area in Sq. Ft.; C=0.62; $\sqrt{2g}=8.02$

g=Velocity of a falling body at the end of one second.

One Miner's Inch=1/50 Second Feet.

Effective Head		Discharge in Miner's Inches								
in Feet	in Ins.	Area of Orifice in Square Feet								
		¼	½	¾	1	1½	2	3	4	5
.00	0	0	0	0	0	0	0	0	0	0
.01	⅛	6.2	12.4	18.7	24.9	37.4	49.8	74.7	99.6	124.5
.02	¼	8.8	17.6	26.4	35.2	52.8	70.4	105.6	140.8	176.0
.03	⅜	10.8	21.5	32.2	43.0	64.5	86.0	129.0	172.0	215.0
.04	½	12.4	24.8	37.3	49.7	74.6	99.4	149.1	198.8	248.5
.05	⅝	13.9	27.8	41.7	55.6	83.4	111.2	166.8	222.4	278.0
.06	¾	15.2	30.4	45.7	60.9	91.4	121.8	182.7	243.6	304.5
.07	⅞	16.4	32.9	49.4	65.8	98.7	131.6	197.4	263.2	329.0
.08	1	17.6	35.1	52.7	70.3	195.4	140.6	210.9	281.2	351.5
.09	⅛	18.6	37.3	56.0	74.6	111.9	149.2	223.8	298.4	373.0
.10	¼	19.6	39.3	59.0	78.6	117.9	157.2	235.8	314.4	393.0
.11	⅜	20.6	41.2	61.9	82.5	123.8	165.0	247.5	330.0	412.5
.12	½	21.5	43.0	64.6	86.1	129.2	172.2	258.3	344.4	430.5
.13	9/16	22.4	44.8	67.2	89.6	134.4	179.2	268.8	358.4	448.0
.14	⅝	23.2	46.5	69.8	93.0	139.5	186.0	279.0	372.0	465.0
.15	¾	24.1	48.2	72.2	96.3	144.4	192.6	288.9	385.2	481.5
.16	⅞	24.8	49.7	74.6	99.4	149.1	198.8	298.2	397.6	497.0
.17	2	25.6	51.2	76.9	102.5	153.8	205.0	307.5	410.0	512.5
.18	⅛	26.4	52.8	79.1	105.5	158.2	211.0	316.5	422.0	527.5
.19	¼	27.1	54.2	81.3	108.4	162.6	216.8	325.2	433.6	542.0
.20	⅜	27.8	55.6	83.4	111.2	166.8	222.4	333.6	444.8	556.0
.21	½	28.5	57.0	85.4	113.9	170.8	227.8	341.7	455.6	569.5
.22	⅝	29.2	58.3	87.4	116.6	174.9	233.2	349.8	466.4	583.0
.23	¾	29.8	59.6	89.4	119.2	178.8	238.4	357.6	476.8	596.0
.24	⅞	30.4	60.9	91.4	121.8	182.7	243.6	365.4	487.2	609.0
.25	3	31.1	62.2	93.2	124.3	186.4	248.6	372.9	497.2	621.5
.26	⅛	31.7	63.4	95.1	126.8	190.2	253.6	380.4	507.2	634.0
.27	¼	32.3	64.6	96.9	129.2	193.8	258.4	387.6	516.8	646.0
.28	⅜	32.9	65.8	98.7	131.6	197.4	263.2	394.8	526.4	658.0
.29	½	33.5	67.0	100.4	133.9	200.8	267.8	401.7	535.6	669.5
.30	⅝	34.0	68.1	102.2	136.2	204.3	272.4	408.6	544.8	681.0
.31	¾	34.6	69.2	103.8	138.4	207.6	276.8	415.2	553.6	692.0
.32	⅞	35.2	70.3	105.4	140.6	210.9	281.2	421.8	562.4	703.0
.33	4	35.9	71.8	107.6	143.5	215.2	287.0	430.5	574.0	717.5
.34	⅛	36.2	72.5	108.8	145.0	217.5	290.0	435.0	580.0	725.0
.35	¼	36.8	73.6	110.3	147.1	220.6	294.2	441.3	588.4	735.5
.36	⅜	37.3	74.6	111.9	149.2	223.8	298.4	447.6	596.8	746.0
.37	½	37.8	75.6	113.4	151.2	226.8	302.4	453.6	604.8	756.0
.38	9/16	38.3	76.6	115.0	153.3	230.0	306.6	459.9	613.2	766.5
.39	⅝	38.8	77.6	116.4	155.2	232.8	310.4	465.6	620.8	776.0
.40	¾	39.3	78.6	117.9	157.2	235.8	314.4	471.6	628.8	786.0
.41	⅞	39.8	79.6	119.4	159.2	238.8	318.4	477.6	636.8	796.0
.42	5	40.3	80.5	120.8	161.1	241.6	322.2	483.3	644.4	805.5
.43	⅛	40.8	81.5	122.2	163.0	244.5	326.0	489.0	652.0	815.0
.44	¼	41.2	82.4	123.7	164.9	247.4	329.8	494.7	659.6	824.5
.45	⅜	41.7	83.4	125.1	166.8	250.2	333.6	500.4	667.2	834.0
.46	½	42.2	84.3	126.4	168.6	252.9	337.2	505.8	674.4	843.0
.47	⅝	42.6	85.2	127.8	170.4	255.6	340.8	511.2	681.6	852.0
.48	¾	43.0	86.1	129.2	172.2	258.3	344.4	516.6	688.8	861.0
.49	⅞	43.5	87.0	130.5	174.0	261.0	348.0	522.0	696.0	870.0
.50	6	44.0	87.9	131.8	175.8	263.7	351.6	527.4	703.2	879.0

Table No. VII. (Continued)

DISCHARGE TABLE FOR STANDARD SUBMERGED ORIFICE

Discharge $=AC\sqrt{2gh}$; A=Area in Sq. Ft.; C=0.62; $\sqrt{2g}=8.02$

g=Velocity of a falling body at the end of one second

One Miner's Inch=1/50 Second Feet.

Effective Head		Discharge in Miner's Inches								
Feet in	Ins. in	Area of Orifice in Square Feet								
		¼	½	¾	1	1½	2	3	4	5
.51	6 ⅛	44.4	88.8	133.1	177.5	266.2	355.0	532.5	710.0	887.5
.52	¼	44.8	89.6	134.5	179.3	268.9	358.6	537.9	717.6	896.5
.53	⅜	45.2	90.5	135.8	181.0	271.5	362.0	543.0	724.0	905.0
.54	½	45.7	91.4	137.0	182.7	274.0	365.4	548.1	730.8	913.5
.55	⅝	46.1	92.2	138.3	184.4	276.6	368.8	553.2	737.6	922.0
.56	¾	46.5	93.0	139.5	186.0	279.0	372.0	558.0	744.0	930.0
.57	⅞	46.9	93.8	140.8	187.7	281.6	375.4	563.1	750.8	938.5
.58	7	47.3	94.6	142.0	189.3	284.0	378.6	567.9	757.2	946.5
.59	⅛	47.8	95.5	143.2	191.0	286.5	382.0	573.0	764.0	955.0
.60	¼	48.2	96.3	144.4	192.6	288.9	385.2	577.8	770.4	963.0
.61	⅜	48.6	97.1	145.6	194.2	291.3	388.4	582.6	776.8	971.0
.62	½	49.0	97.9	146.8	195.8	293.7	391.6	587.4	783.2	979.0
.63	9/16	49.3	98.6	148.0	197.3	296.0	394.6	591.9	789.2	986.5
.64	⅝	49.7	99.4	149.2	198.9	298.4	397.8	596.7	795.6	994.5
.65	¾	50.1	100.0	150.3	200.4	300.6	400.8	601.2	801.6	1002.0
.66	⅞	50.6	101.2	151.7	202.3	303.4	404.6	606.9	809.2	1012.0
.67	8	50.9	101.8	152.6	203.5	305.2	407.0	610.5	814.0	1018.0
.69	¼	51.6	103.2	154.9	206.5	309.8	413.0	619.5	826.0	1032.0
.71	½	52.4	104.8	157.1	209.5	314.2	419.0	628.5	838.0	1048.0
.73	¾	53.1	106.2	159.3	212.4	318.6	424.8	637.2	849.6	1062.0
.75	9	53.8	107.6	161.5	215.3	323.0	430.6	645.9	861.2	1076.0
.80	⅝	55.6	111.2	166.8	222.4	333.6	444.8	667.2	889.6	1112.0
.85	10 ¼	57.3	114.6	171.9	229.2	343.8	458.4	687.6	916.8	1146.0
.90	¾	59.0	117.9	176.8	235.8	353.7	471.6	707.4	943.2	1179.0
.95	11 ⅜	60.6	121.2	181.7	242.3	363.4	484.6	726.9	969.2	1212.0
1.00	12	62.2	124.3	186.4	248.6	372.9	497.2	745.8	994.4	1243.0
1.05	12 ⅝	63.7	127.4	191.1	254.8	382.2	509.6	764.4	1019.0	1274.0
1.10	13 ¼	65.2	130.4	195.6	260.8	391.2	521.6	782.4	1043.0	1304.0
1.15	13 ¾	66.6	133.3	200.0	266.6	400.0	533.2	799.8	1066.0	1333.0
1.20	14 ⅜	68.1	136.2	204.2	272.3	408.4	544.6	816.9	1089.0	1362.0
1.25	15	69.5	139.0	208.5	278.0	417.0	556.0	834.0	1112.0	1390.0
1.30	⅝	70.9	141.8	212.6	283.5	425.2	567.0	850.5	1134.0	1418.0
1.35	16 ¼	72.2	144.4	216.6	288.8	433.2	577.6	866.4	1155.0	1444.0
1.40	16 ¾	73.6	147.1	220.6	294.2	441.3	588.4	882.6	1177.0	1471.0
1.45	17 ⅜	74.8	149.7	224.6	299.4	449.1	598.8	898.2	1198.0	1497.0
1.50	18	76.1	152.2	228.4	304.5	456.8	609.0	913.5	1218.0	1522.0
1.60	19 ¼	78.6	157.2	235.9	314.5	471.8	629.0	943.5	1258.0	1572.0
1.70	20 ⅜	81.0	162.1	243.2	324.2	486.3	648.4	972.6	1297.0	1621.0
1.80	21 ⅝	83.4	166.8	250.2	333.6	500.4	667.2	1001.0	1334.0	1668.0
1.90	22 ¾	85.7	171.4	257.0	342.7	514.0	685.4	1028.0	1371.0	1714.0
2.00	24	87.9	175.8	263.7	351.6	527.4	703.2	1055.0	1406.0	1758.0
3.00	36	107.8	215.5	323.2	431.0	646.5	862.0	1293.0	1724.0	2155.0
4.00	48	124.0	248.0	372.0	496.0	744.0	992.0	1488.0	1984.0	2480.0
5.00	60	138.8	277.5	416.2	555.0	832.5	1110.0	1665.0	2220.0	2775.0
6.00	72	152.0	304.0	456.0	608.0	912.0	1216.0	1824.0	2432.0	3040.0
7.00	84	164.0	328.5	492.8	657.0	986.0	1314.0	1971.0	2628.0	3285.0
8.00	96	176.0	352.0	528.0	704.0	1056.0	1408.0	2112.0	2816.0	3520.0
9.00	108	186.8	373.5	560.2	747.0	1120.0	1494.0	2241.0	2988.0	3735.0
10.00	120	196.8	393.5	590.2	787.0	1180.0	1574.0	2361.0	3148.0	3935.0

The effective head is the difference between the water level above the orifice and the water level below the orifice. When the upper and lower gages are set on the same level it is equal to the difference in their readings.